Isambard Kingdom Brunel

An illustrated life of Isambard Kingdom Brunel

1806-1859

Richard Tames

Shire Publications Ltd

Isambard Kingdom Brunel

A silhouette of Sir Marc Isambard Brunel, father of Isambard Kingdom Brunel.

Copyright © 1972 by Richard Tames. First published 1972. Reprinted 1975, 1977, 1980, 1982, 1983 and 1985. ISBN 0 85263 140 5.
All rights reserved. No part of this publication may be reproduced or transmitted in any form or by any means, electronic or mechanical, including photocopy, recording, or any information storage and retrieval system, without permission in writing from the publishers, Shire Publications Ltd, Cromwell House, Church Street, Princes Risborough, Aylesbury, Bucks, HP17 9AJ, U.K.

Printed by C. I. Thomas & Sons (Haverfordwest) Ltd.

CONTENTS

ACKNOWLEDGEMENTS

The author wishes to thank the following for permission to reproduce the illustrations on the pages indicated: Robert D. Bristow, 44; British Rail (Western Region), 26; The Brunel Society, 25; R. A. Buchanan, Centre for the Study of the History of Technology, Bath University, 16; The City Museum and Art Gallery, Bristol (Curator of Technology: Paul W. Elkin), 10, 22, 32, 33, 34; Lady Cynthia Gladwyn, 2; Illustrated London News, 39; R. Lea, 20; The Museum of British Transport, Clapham, 4, 21, 24; Radio Times Hulton Picture Library, 28, 37, 41; The Science Museum, London, 30; The Smithsonian Institution, Washington D.C., 6; University of Bristol Arts Faculty Photographic Unit, 19; Reece Winstone, 11, 12, 13, 35.

EARLY LIFE

Father and son

So great is the fame of Isambard Kingdom Brunel that it has almost totally eclipsed the reputation of his father, Marc Brunel, to whom he owed so much. Any consideration of the career of the son must begin by acknowledging this debt. Their talents in many ways were similar. Both were men of outstanding imagination and ingenuity, both had the capacity to turn visions into realities through hard work and incredible application to detail. But whereas the father lived to a peaceful old age, the son burned out his energies to find a lasting reputation and a premature grave. Whereas the father was simple, unassuming and absent-minded, the son was complex and sophisticated, a shrewd judge of men, with a withering tongue and commanding presence in spite of his shortness of stature. Not that Isambard was unaware of his shortcomings, confiding to his journal that 'my self-conceit and love of glory vie with each other which shall govern me.'

Even as an infant Isambard (Kingdom was his mother's surname) displayed a precocious talent for drawing and by the age of six had mastered the principles of geometry. His delighted father sent him to boarding school in Hove, where he amused himself by making a survey of the town and sketching its more significant buildings. At the age of fourteen he was sent to the College of Caen in his father's native Normandy and thence to the Lycée Henri-Quatre in Paris, then renowned for its teachers of mathematics. Having thus attended to the theoretical aspects of his son's apprenticeship as an engineer, Marc had him serve under the master-craftsman Louis Breguet, who instructed him in the practical arts of making watches and scientific instruments. At the age of sixteen, in 1822, Isambard returned to England to start work in his father's office at 29 Poultry, continuing his practical education by almost daily visits to that great school of engineers, Maudslay, Sons and Field of Lambeth.

Left: Isambard Kingdom Brunel, from an engraving.

His natural talents thus carefully nurtured by the most suitable training his father could contrive, Isambard was ready, at eighteen years of age, to assist him in the greatest engineering project of his father's distinguished career—the boring of a tunnel under the Thames.

The Thames tunnel

A Thames tunnel had first been suggested in 1798 and a Thames Archway Company was incorporated in 1805 to construct it. The original engineer, Vazie, having exhausted the initial capital without making substantial progress, was obliged to hand over to the famed Richard Trevithick, inventor of the first successful locomotive and pioneer of high-pressure steam-engines. Working with a picked team of Cornish miners Trevithick overcame the series of treacherous quicksands which had defeated his predecessor and drove 1000 feet of tunnel (out of a total length of 1200 feet) in just six months. An abnormally

Sir Marc Brunel's tunnelling shield, inspired by his observation of the ship-worm, began work on the Thames tunnel in 1825.

high tide in January 1808 overwhelmed the pumps, however, and Trevithick and his men were lucky to escape with their lives.

Fifteen years after this debacle, Marc Brunel took up the challenge, armed with a new device of his own invention, a tunnelling shield, the principle of which had been suggested to him by his observation of the destructive 'ship-worm' while working at Chatham Dockyard. The site selected for this new attempt was three quarters of a mile west of Trevithick's abandoned workings, running from Rotherhithe to Wapping. Having sunk a great shaft 50 feet across and 40 feet deep, Brunel senior ordered the massive cast iron frame to begin edging forward beneath the Thames on 28th November 1825. It soon became apparent that the geologist's promise of solid clay was false and water burst into the workings periodically, reducing progress to a crawl. The directors, overriding Brunel's warnings, put the men on piece rates to encourage them to even greater efforts. Incentives were certainly needed, the Thames serving London as an open sewer and consequently afflicting the labourers with its stench and a 'tunnel sickness', which struck men blind suddenly and permanently. William Armstrong, the resident engineer, broke down under the strain in April 1826, to be replaced by young Isambard, then barely twenty, but already more than equal to the task. Cat-napping in the tunnel itself, Isambard frequently worked below for 36 hours at a stretch. He revelled in the work, despite the irritations occasioned by the shilling-a-visit sightseers who came to see engineering history being made. Brunel senior, however, was increasingly oppressed by a sense of impending disaster and on 13th May 1827 recorded in his diary that 'during the preceding night the whole of the ground over our heads must have been in movement . . . The shield must, therefore, have supported upwards of 600 tons . . . Notwithstanding every prudence on our part a disaster may still occur. May it not be when the arch is full of visitors!'

Five days later Brunel senior's worst fears were realised. Leakages at high water became uncontrollable and a massive tidal wave swept through the tunnel, carrying all before it. Isambard managed to collect his men and get them half-way up the shaft before the water swept away the lower section of the stairway. When they reached the surface, however, a faint cry was heard from below and without a moment's hesitation Brunel slid down one of the iron ties of the shaft and tied a rope round the half-conscious navvy. They were hauled to safety and the roll-call revealed that not a single man had been lost.

The following day, as the curate of Rotherhithe warned his congregation that the disaster was 'but a just judgement upon the presumptuous aspirations of mortal men', Isambard was already working on the river bed, inspecting the damage from a diving bell. By 11th June nearly 20,000 cubic feet of clay had been flung into the hole, and the pumps had cleared the shaft and the first 150 feet of the tunnel by the 25th. By November 1827 the whole tunnel had been cleared and Brunel resolved to celebrate by holding a banquet in the reclaimed workings. Hung with crimson drapery and lit by gas candelabra, the tunnel was an impressive setting for the select company of 50 who sat with Brunel and the 120 elite miners who feasted not far away, all dining to the accompaniment of the band of the Coldstream Guards.

The celebration was, however, premature. On 12th January disaster struck again with renewed force. Once again the water overwhelmed the shield and Brunel was saved only by the tidal wave itself, which, having knocked him senseless, bore him along the tunnel and up one of the shafts, where his inert form was snatched from the murky tide. His reminiscence of this close escape reveals his calm detachment in the face of hazard and danger–'When we were obliged to run I felt nothing in particular; I was only thinking of the best way of getting us out and the probable state of the arches. When knocked down, I certainly gave myself up, but I took it very much as a matter of course . . . for I never expected we should get out.' All he remembered of the rushing water was that the effect 'was grand, *very grand*' and concluded that 'the sight and whole affair was well worth the risk and I would willingly pay my share . . . of the expenses of such a "spectacle"'. As soon as he was conscious he directed diving operations to check the extent of the damage. Not until this procedure was completed did he consent to be carried home, where a broken leg obliged him to rest for many months. Work on the tunnel ceased for seven years and, when it was finally resumed, Isambard Brunel would be far too occupied with even greater things to lend a hand. But first there were years of frustration to be endured.

A mediocre success

As Brunel lay convalescing he dreamed of the great things he would achieve–build a fleet of ships and storm Algiers, construct a new London Bridge with an arch of 300 foot span, build tunnels at Gravesend and Liverpool and 'at last be rich, have a house built, of which I have even made the *drawings* etc., be the first engineer and an example for future ones'. The thought that so far his achievements consisted of one abandoned tunnel brought him back to earth and he was also moved to contemplate a more barren future in which he might be 'unemployed, untalked of, *penniless*'. Worse still, he might be neither— 'I suppose a sort of middle path will be the most likely one—a mediocre success—an engineer sometimes employed, sometimes not—£200 or £300 a year and that uncertain'.

In the meantime he would try his hand at as many projects as possible, in the hope that one would turn up trumps. The most promising was his 'gaz engine', a device intended to supersede steam-power by using the gas generated by carbonate of ammonia and sulphuric acid. The Admiralty itself expressed interest and even helped to subsidize the experiments, conducted on the abandoned site at Rotherhithe. More than four years passed before the dogged Brunel admitted defeat, but in January 1833 he was forced to admit 'all the time and expense, both enormous . . . are . . . wasted—It must therefore die and with it all my fine hopes—crash—gone—well, well, it can't be helped'.

The 'gaz engine' was, however, by no means his only project, and with the help of his father and friends he acquired a number of useful commissions—drainage works on the Essex coast at Tollesbury, a new dock at Monkwearmouth (Sunderland) and numerous surveys of canals, waterways and bridges. All this meant ceaseless travel and with it the opportunity to broaden his knowledge by examining sites and buildings all over the country. Nor did he cease dreaming. While riding on the newly opened Liverpool and Manchester Railway he wrote prophetically, 'I record this specimen of the shaking of Manchester railway. The time is not far off when we shall be able to take our coffee and write while going noiselessly and smoothly at 45 m.p.h.—let me try'.

The outlook was still far from promising. Work on Monkwearmouth dock was suspended indefinitely and so was the proposed new dry dock at Woolwich, on which he had lavished many hours in surveys and trial borings. A minor triumph, an observatory at Kensington for Sir James South, was marred by a squalid wrangle over payment and his applications for the post of engineer to the Newcastle and Carlisle Railway and the Bristol and Birmingham Railway were both turned down.

The Clifton bridge

His chance came, at last, while he was still convalescing, with the announcement of a competition to design a bridge to cross the river Avon at Bristol, a stupendous challenge which involved spanning the mighty Avon gorge. Brunel decided on a suspension bridge and submitted four designs for bridges situated at four possible sites along the gorge. Bold and imaginative in conception they were presented with the most exquisite draughtsmanship of which he was capable. Thomas Telford, the aged first President of the Institute of Civil Engineers, acted as judge

This drawing by Brunel, dated 1831, was probably one of the designs which he submitted to the second competition for the Clifton Bridge. A similar sketch, showing the Egyptian-style decoration which was finally adopted, has not survived.

The Clifton bridge

This photograph of Brunel was probably taken in the 1850s.

The Clifton Suspension Bridge, designed by Isambard Brunel. Above left: an 1836 print of the design which won the competition, showing sphinxes on top of the towers. Above right: construction was abandoned in the 1830s and not resumed until 1861 after Brunel had died. The two towers

on either side of the Avon gorge became known as the 'Follies'. Below left: the two sides of the road about to meet as the bridge nears completion in 1864. Below right: Clifton Bridge as it is today. The bridge is 230 feet above high water, has a span of 630 feet and weighs 7000 tons.

on behalf of the Bridge Committee and rejected every single entry submitted.

Nonplussed, the Committee commissioned a design from the master himself. It was, in the words of an eminent historian 'the one truly monstrous aberration of his long career', and a second competition was announced. This time Brunel won but success was once more dashed from his grasp as a local political convulsion overwhelmed the bustling life of Bristol. 1830 had been a long hot summer of riots and rick-burning over the south of England; in the Midlands and North demands for parliamentary reform revived and by the following year a full-scale campaign of agitation was under way. When the Lords rejected a Reform Bill which the Commons had passed, Bristol exploded in three days of rioting and arson. Brunel, although sympathetic to reform, served as a special constable to restore law and order. Armed with the back of a broken chair he helped salvage the corporation plate from the deserted Mansion House and subsequently gave evidence at the trial of the luckless mayor. With the city half devastated the bridge scheme joined Brunel's other commissions at Monkwearmouth and Woolwich and was shelved indefinitely. Work did not start till 1835, and after further delay, was not completed until 1864.

The effort had not been wasted, however, for Brunel soon found himself involved in the more mundane task of scouring out Bristol Docks, which were constantly silting up and it was this work which brought him into contact with a group of promoters who were planning a railway from Bristol to London and were looking for a surveyor. Brunel impressed them with his energy and vision, but declined to give his assent to any scheme except that which would authorise not the cheapest route, but the best. Doubly impressed by this risk-all display of professional integrity the Railway Committee appointed him engineer. It was the turning-point of his career.

THE GREAT WESTERN RAILWAY

Engineer to the GWR

Appointed in March 1833, Brunel was ordered to complete his preliminary survey by May. Hectic, sleepless days of work and constant travel by horse and coach enabled him to meet his deadline, as well as continue to supervise the Bristol dock scheme and a survey of the Fossdyke navigation. To carry him about his business he designed his famous 'Flying Hearse', a black britzska, which held his plans, engineering instruments, a monster case of fifty cigars and a seat which extended into a couch to enable him to snatch a few hours sleep whenever chance arose. Even the demonic Brunel could scarcely sustain the pace, however, and confessed to his assistant that, 'between ourselves it is harder work than I like. I am rarely much under 20 hours a day at it.'

Final plans were completed by November and in March 1834 the bill needed to incorporate the company and grant it powers of compulsory purchase was referred to a committee of the Commons. Here rival interests—landowners, coach proprietors, the Kennet and Avon Canal Company, the London and Windsor Railway and the London and Southampton Railway—united to defeat the proposed Great Western Railway. An epic parliamentary battle lasted for 57 days and ended in the rejection of the bill by the Lords. Undismayed, the Great Western directors submitted a fresh bill the following year and thrust the young Brunel in before the Commons committee to defend their case. His cross-examination lasted eleven days. According to an eyewitness 'his knowledge of the country surveyed by him was marvellously great . . . He was rapid in thought, clear in his language and never said too much or lost his presence of mind. I do not remember ever having enjoyed so great an intellectual treat as that of listening to Brunel's examination.' With the support of such eminent engineers as Palmer, Locke, Vignoles and the great George Stephenson himself the forty day contest was finally decided in favour of the GWR, but the cost was nearly £90,000 in legal fees and 'parliamentary expenses'. Brunel, typically, was already engaged in marking out new

15

Brunel's lock at Bristol seen at low tide. One of the two caisson gates closed against the sill in the masonry and was swung open into the recess on the left.

routes, to Cheltenham, Gloucester, Worcester, Oxford, Exeter, Plymouth and South Wales, making the GWR itself the spine of a mighty West of England railway empire.

On Boxing Night 1835 Isambard Brunel sat late and alone in his office in Parliament Street, and, taking up the diary which two years of ceaseless work had forced him to abandon, set down his thoughts:

'What a blank in my journal! And during the most eventful part in my life. When I last wrote in this book I was just emerging from obscurity. I had been toiling most unprofitably at numerous things—unprofitably at least at the moment. The Railway certainly was brightening but still very uncertain—what a change. *The Railway* is now in progress. I am their Engineer to the finest work in England—a handsome salary—£2000 a year—on excellent terms with my Directors and all going smoothly, but what a fight we have had—and how near defeat— and what a ruinous defeat it would have been. . . . And it's not this alone but everything I have been engaged in has been successful.

'*Clifton Bridge*—my first child, my darling, is actually going—recommenced week last Monday—Glorious!

'*Sunderland Docks* too going well.

'*Bristol Docks* all Bristol is alive and turned bold and speculative with this Railway—we are to widen the entrances and the Lord knows what.'

Four branch lines were added to the list and a suspension bridge across the Thames—'I have condescended to be engineer to this—but I shan't give myself much trouble about it. If done, however, it all adds to my stock of irons.'

'I think this forms a pretty list of real profitable, sound professional jobs—unsought for on my part, that is given to me fairly by the respective parties, all, except MD [Monkwearmouth Dock] resulting from the Clifton Bridge—which I fought hard for and gained only by persevering struggles and some manoeuvres (all fair and honest however). Voyons.

'I forgot also Bristol and Gloster Railway.' All told, these projects represented capital of more than £5,300,000, say ten times as much in modern terms—'I really can hardly believe it when I think of it', he concluded in self-amazement, and then finished the entry with a warning note to himself. 'Everything has prospered everything at this moment is sunshine. I don't like it—it can't last—bad weather must surely come. Let me see the storm in time to gather in my sails.'

Private life

Only the exceptional man can combine devotion to his work with devotion to a wife and family. Marc Brunel was one of those men; his son was not. Having 'arrived' as an engineer Isambard determined that 'this time 12 months I shall be a married man'. His diary recorded not only the resolution but the doubt—'How will that be? Will it make me happier?' In the event, it made little difference. The beautiful and gifted Mary Horsley, whom Brunel had known for more than five years, consented to become his wife and they were married on 5th July 1836. After a brief honeymoon they settled into No. 18, Duke Street, Westminster, which was to become their home for the whole of their marriage, a marriage of conventional and, it seems, fairly superficial, happiness. Isambard continued to regard his work as his real 'wife' and Mary was free to play the *grande dame,* spending, dressing, riding, entertaining. She rarely accompanied her husband in his travels, but was usually present at the ceremonies which marked the inauguration or completion of one of his projects. They had three children—Isambard, who became a lawyer, to his father's mortification; Henry Marc, who did become an engineer, and Florence Mary, who married a master at Eton College.

Brunel, who was very fond of children, had a genius for inventing games and doing tricks, one of which nearly cost him his life. Having palmed a half-sovereign to make it reappear from his nose or mouth, he accidentally swallowed the coin, which lodged in his throat and threatened to choke him to death. An emergency tracheotomy with a two-foot long pair of forceps failed to release the obstruction. Brunel, having suffered fruitlessly at the hands of the most eminent surgeons of the day, found his own salvation by designing a simple device to dislodge the coin by centrifugal force. Consisting essentially of a board between two upright pivots, the apparatus was to whirl him head over heels until the sovereign came out of its own accord. The first trial reduced him to a coughing fit which his friends feared would be fatal, but he ordered a second attempt to be made. The coin dropped out and Brunel was restored to his feet and his health, unharmed.

The finest work in England

When Brunel began work on the GWR he was but 30 years of age, with no previous railway experience and no trained assistants. The best railway contractors and foremen were occupied on the North-South link via Birmingham and so Brunel was faced with the dual problem of supervising the greatest construction project in recorded history and simultaneously welding his own work force into a disciplined and efficient team. He was not a man to suffer fools gladly and the following letter to an erring draughtsman may perhaps be taken as an example of his attitude to those who let him down:

'Plain, gentlemanly language seems to have no effect upon you. I must try stronger language and stronger measures. You are a cursed, lazy, inattentive, apathetic vagabond, and if you continue to neglect my instructions, and to show such infernal laziness, I shall send you about your business. I have frequently told you, amongst other absurd, untidy habits, that that of making drawings on the back of others was inconvenient; by your cursed neglect of that you have again wasted more of my time than your whole life is worth, in looking for the altered drawings you were to make of the Station—they won't do!'

Brunel was determined to elevate railway travelling to an altogether higher plane of smoothness and comfort. The route he had surveyed was unparalleled for the infrequency and gradual nature of its gradients. Rejecting Stephenson's 4 foot 8½ inch gauge (derived empirically from the average of the coal carts traditionally employed on Tyneside), Brunel set himself to work out from first principles the gauge most suitable for speed and smoothness of operation. Calculation and experiment led him to adopt 7 feet 0½ inch which would, he argued, accommodate more powerful engines, and larger, more stable carriages, travelling regularly at unprecedented speeds. This 'broad gauge' was, without doubt, eminently suitable for the flat, straight run from London to Bristol but Brunel chose to play down the problems of increased land and costs for curves and points and earthworks and to ignore the problem of what would happen when the two gauges met. This was all in the future, although every engineer in England felt it necessary to declare himself for one side or the other, and in the meantime Brunel busied himself with supervising the numerous subcontractors under his charge. His notebooks are an incredible *pot-pourri* of facts and figures which reveal his obsessive concern for detail—tables of rainfall figures, local times along the route (GMT had not yet

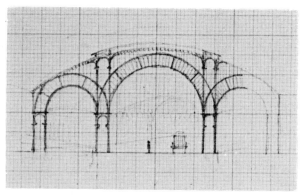

From Brunel's sketchbooks in Bristol University Library, this is a drawing for Paddington station.

The sounding arch of Brunel's bridge taking the Great Western Railway over the Thames at Maidenhead.

been adopted over the whole country) and even notes on the best species of grass for consolidating new cuttings and embankments.

The first completed section of rail, from Paddington to Taplow (Maidenhead) (Brunel's bridge over the Thames here had the world's largest brick-built span, a record that still stands today) was opened on 4th June 1838 but Brunel scarcely had time to congratulate himself upon the progress made. He confided to a friend that, 'if ever I go mad, I shall have the ghost of the opening of the railway . . . standing in front of me, holding out its hand, and when it steps forward, a little swarm of devils in the shape of leaky tanks, uncut timber, half-finished station houses, sinking embankments, broken screws, absent guard plates, unfinished drawings and sketches will . . . lift up my ghost and put him a little further off than before.' Nor were human obstacles lacking. A number of north of England shareholders, prejudiced against this presumptuous, half-French Londoner, began to agitate for a joint engineer to be appointed to share the work with him. A couple of derailments and the bumpy rides experienced on the first stretch of line seemed to add weight to their arguments.

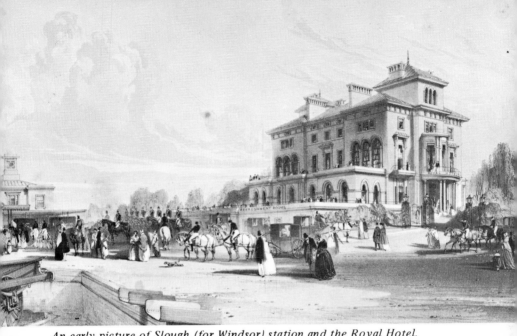

An early picture of Slough (for Windsor) station and the Royal Hotel. Queen Victoria took her first train ride from here in 1842.

Brunel was able to exonerate himself by proving that the rolling-stock, designed by traditional coach-builders, not engineers, was primarily to blame for the bumpy ride. The appointment of the talented Daniel Gooch as his Chief Locomotive Assistant enabled him to overcome problems of motive power but he still found himself in trouble at the half-yearly meeting held in August 1838, when two engineers, Hawkshaw and Wood, were invited to report on the progress of the line. Hawkshaw's report was lightweight stuff and Brunel soon demolished it. Wood, however, came up with an apparently logical explanation for the poor performance of Brunel's locomotives at high speed. It was, he alleged, due to wind resistance caused by the increased frontal area of the broad-gauge engines. Brunel, with Gooch's able assistance, began a series of trials to disprove this argument and soon found the real cause of the trouble—a too-narrow blast pipe which throttled the engine at high speeds. Minor adjustments enabled Brunel to demonstrate triumphantly to the directors that his engines could not only pull heavy trains at more than 40 m.p.h., they could also do so on less fuel than anyone had ever thought possible.

Vindicated against his human opponents, he then turned to

21

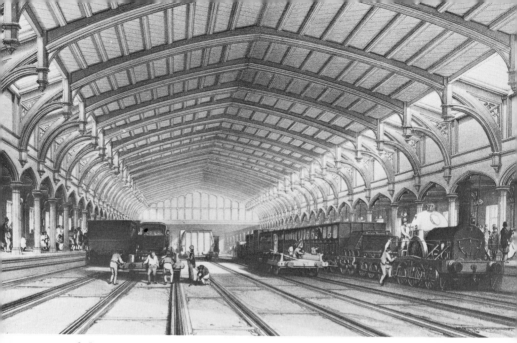

This lithograph of the interior of Bristol station (Temple Meads) was one of a series by J. C. Bourne published in 1842 and showing the GWR in its original form. Trains first ran from Bristol to Bath on 31st August 1840, and to London on 30th June 1841. Note the broad-gauge locomotive and rolling stock.

do battle with the elements. Torrential rains in the autumn of 1839 had drowned the Thames Valley, reducing the two mile Sonning Hill cutting to a quagmire. Brunel assumed personal control of the 1200 navvies and 200 horses on this sector and completed the cutting by the year's end. On 17th March 1840 the line was opened as far as Reading and its smoothness was convincingly demonstrated by Gooch's new *Firefly* which *averaged* 50 m.p.h. on a 30 mile run.

Work on the Bristol end proceeded more slowly, owing to the much larger number of difficult projects involved. The entire terminus at Temple Meads, complete with a vast hammer-beam roof, was built on arches 15 feet above ground level. Bath station, similarly elevated, was approached by a viaduct of 73 arches and between the two stations there was another viaduct of 28 arches, four major bridges and seven tunnels. Nevertheless this section was opened for traffic on 31st August 1840.

By May 1841 only the Chippenham-Bath section remained incomplete. It bristled with difficulties—a crossing of the Avon,

The west front of Box Tunnel drawn by Bourne. Nearly 2 miles long, far greater than any previous tunnel, Box cost the lives of more than a hundred men.

viaducts at both ends of the section, the diversion of the Kennet and Avon canal and finally what his critics called the 'monstrous and extraordinary, most dangerous and impracticable tunnel at Box'. Nearly two miles long, it was by far the greatest tunnel ever attempted and cost the lives of more than 100 men in the building. For two and a half years this single project accounted for a ton of gunpowder and a ton of candles every week. 100 horses and carts brought in 30 million bricks and on two occasions work was brought to a complete standstill by flooding. In December 1840, four months after the tunnel should have been completed, Brunel took personal charge of the 4,000 men and 300 horses employed round the clock to complete one of the greatest projects since the Pyramids. The work was finished by June 1841 and the whole way open from London to Bristol. It had cost £6½ million, more than double the original estimate, but it was, without doubt, 'the finest work in England'.

The battle of the gauges
   As Brunel had envisaged, the GWR sprouted branches which

reached out to Exeter, Oxford and Cheltenham. A time of four hours forty minutes for the Exeter-Paddington run was hailed as a record high-speed performance and, to its supporters, convincing proof of the superiority of broad gauge. Not so its critics and when the narrow gauge met its rival at Gloucester, parliament appointed a Royal Commission to investigate the matter. Brunel proposed to resolve the issue by speed trials, a sporting gesture in view of the fact that a number of improved locomotives had just come into service on the narrow gauge. In mid-December 1845 Gooch's 'Firefly' class *Ixion* made three round trips from Paddington to Didcot (106 miles total) with

The break of gauge at Gloucester. Despite the superior performance of Brunel's broad-gauge trains, Parliament declared the 4ft. 8½in. gauge to be standard and the GWR had to change.

loads of 80, 70 and 60 tons, averaging more than 50 m.p.h. with the 60-ton load and touching 60 with her maximum load. Brunel's northern rivals, running from York to Darlington and back (88 miles) could only achieve 53 m.p.h. with a 50-ton load with their newest locomotive, while the tried and tested *Stephenson* ran off the rails and turned over halfway out on the first leg.

The Gauge Commissioners were less exhilarated by Brunel's victory than they were sobered by the thought that speed was not everything and the narrow gauge was far more economic to construct. Besides it was easier and cheaper to narrow the gauge of a railway than widen it and whereas only 274 miles of broad-gauge track was in operation, narrow-gauge working already totalled 1,900 miles. Narrow gauge therefore became 'standard' for the whole country and the GWR was obliged to lay a third intermediate rail to afford facilities for narrow-gauge traffic. Pamphlet warfare between the two sides continued intermittently for some time and Brunel toyed with a project for 'containerization' to overcome the labour and expense of transhipment from one gauge to the other, but the effect of the Commissioners' decision was to confine the broad gauge to its existing territory. Not until the 1890s, however, did the GWR complete the transition to standard 4 ft. 8½ in. track throughout. Although the 'gauge war' was wasteful and expensive in some ways, it also provided a healthy stimulus to locomotive design and performance which helped Britain establish a technical lead in this field which she was to retain.

The atmospheric railway

Brunel was never satisfied with what had been done and was always looking out for new developments. When the brothers Samuda successfully ran an 'atmospheric railway' one and a quarter miles long on the abandoned roadbed of the Birmingham, Bristol and Thames Junction Railway, Brunel was sure they were on to something promising, though Stephenson dismissed it as 'a great humbug'. In 1843 a slightly longer line from Kingstown (Dublin) to Dalkey was opened to the public and crowds rushed to travel on this silent, smokeless marvel which averaged 30 m.p.h. The principles on which it worked were simple—a pipe was laid between the tracks and within the pipe ran a piston joined to the carriage via a flanged opening along the top of the pipe. Steam-pumps, placed at intervals along the track, extracted the air from the pipe in front of the train and the atmospheric pressure behind the piston then pushed it, and the train, along as it expanded to fill the partial vacuum created by the steam-engine. Simple and elegant, it seemed to Brunel a vision of the future and he dismissed as mere mechanical difficulties the problems of brakes, explosions and leakage of air. In fairness to him it must be said that the Board of Trade and the French Public Works Department were also favourably impressed and Brunel met little opposition when he ordered the installation of the atmospheric system along a sizeable section of the South Devon Railway.

As usual Brunel failed to reckon with the technical limitations

Isambard Kingdom Brunel

Left: old atmospheric railway tubes, photographed about 1911 at Paignton, in use as surface water drains.

Below: the atmospheric railway pumping house at Totnes in 1911. It was never used for its intended purpose.

of his age—the slotted pipes were inaccurately cast, the leather flange was eaten by rats or froze as stiff as a board, the pumping engines were faulty. Nevertheless for several months in 1848 a regular service was operated between Exeter and Newton Abbot and *average* speeds of 64 m.p.h. were attained. Brunel was sure that the installation of electric telegraph communication between the pumping-houses would greatly raise efficiency by enabling the pumps to synchronise their efforts. In February 1848 he reported to the directors that 'we are in a fair way of shortly overcoming the mechanical defects and bringing the whole apparatus into regular and efficient practical working' In June, however, it was discovered that the entire length of the leather valve was disintegrating. To have replaced this alone would have cost £25,000 and Brunel abruptly called off the whole venture, the most costly failure in the history of engineering at that time—the sale of the pumping machinery alone fetching over £40,000.

The Tamar bridge

This disaster was eclipsed, however, by a later triumph as the broad gauge thrust west to Plymouth—a bridge over the Tamar, where the river is 1,100 feet wide and 70 feet deep. Discarding a plan for a timber bridge with six 100-foot spans, and another for a single span of 1,000 feet, Brunel decided on two spans of 465 feet supported by a single deep water pier. The pier was built by using a cast iron cylinder, 35 feet in diameter, as a coffer dam, within which workmen could excavate the mud and sand of the river bed and then erect a masonry column to bear the two massive 1,000 ton trusses. Begun in June 1854 the pier was completed by the end of 1856. 1st September 1857 was the date set for the placing of the first truss which would be floated into position on pontoons and then jacked up to the level of the pier and secured. Before a vast crowd the diminutive figure of Brunel directed his army of workmen by a series of complicated signals, using numbers and flags. The whole operation, on Brunel's insistence, was carried out in complete silence. 'Not a voice was heard', wrote an eye-witness, '. . . as by some mysterious agency, the tube and rail, borne on the pontoons, travelled to their resting place . . . With the impressive silence which is the highest evidence of power, it *slid*, as it were, into its position, without an accident, without any extraordinary mechanical effort, without a 'misfit', to the eighth of an inch'.

The Royal Albert Bridge, Saltash, in 1859, the year of its opening. The two main spans are each of 465 feet.

The tension broke, the Royal Marines played 'See the conquering hero comes' and Brunel, much relieved, stepped down from his platform to acknowledge the ovation of the crowds. The second span at Saltash was floated in July 1858 and in May 1859 the Prince Consort, who had taken a close personal interest in the whole project, himself christened and opened the Royal Albert Bridge.

THE GREAT SHIPS

The 'Great Western'

It was while working on the Great Western Railway that Brunel first became interested in steam navigation. His father had made a number of contributions to the design of marine engines and paddle wheels but believed, as most engineers did, that 'steam cannot do for distant navigation'. The problem was that no steamship could possibly carry enough fuel for an ocean crossing, although engines were useful auxiliaries for an ocean going ship becalmed or fighting contrary winds and currents. Apart from that they appeared to be limited to service as ferries or coasters, regular, reliable and fairly insignificant.

It was Isambard Brunel who found the answer—larger ships. It had always been assumed that if the hull were doubled in size the ship would need double the weight of coal to generate the power necessary to propel it. Brunel spotted the fallacy in the reasoning and formulated a basic rule of naval architecture—that whereas the carrying capacity of a hull increases as the cube of its dimensions, its resistance, that is the power needed to drive it through water, only increases as the square of those dimensions. It was all simply the difference between volume and surface areas and once stated seemed unbelievably simple and obvious. A number of the GWR men were soon won over to the idea of a trans-Atlantic steamship which would make the terminus of their railway not Bristol, but New York and in July 1836 construction of the *Great Western* began.

Built of oak by traditional methods the *Great Western's* only major modification was an immensely strong hull, particularly in the longitudinal plane, to enable her to withstand the worst storms without stress. Within a year she was ready to be towed to London for fitting. To the merchants of London and Liverpool the *Great Western* represented a challenge, a symbolic affirmation of Bristol's determination to regain her eighteenth century ascendancy, and in the two ports the *British Queen* and the *Liverpool* were respectively laid down to join the new ship in a race to New York. When it became obvious that the *Great Western* would be ready long before her rivals the London men chartered the *Sirius* to stand in for the *British Queen*, and on 28th March 1838 she left the Thames to fill her newly enlarged bunkers at Cork, preparatory to attempting the first all-steam crossing of the Atlantic; three days later, after a flurry of final testing, the *Great Western* set out in pursuit.

Near disaster struck the *Great Western* early next morning when the boiler lagging was ignited by sparks from the furnace

A model of the 'Great Western', Brunel's first steamship, which made her first Atlantic crossing in 15 days in 1838.

flues. Within minutes the engine-room was full of smoke and the underside of the deck planking on fire. The engineers displayed great presence of mind and soon had the situation under control, but not before Brunel had, in his haste, plunged nearly 20 feet from a half-burnt ladder, into the water on the engine-room floor. Fortunately one of the men broke his fall and pulled him clear of danger. Brunel was taken by boat to Canvey Island where he remained for several weeks. The race to New York, however, was still on, though it was not until 8th April that the *Great Western* finally left Bristol.

The *Sirius*, a day nearer the destination, had started on the 4th. Running into strong winds, the tiny (703-ton) ship pushed resolutely onwards until, after nineteen days at sea, she reached New York with barely fifteen tons of coal left in her bunkers. A hero's welcome awaited her crew, but they were completely up-staged by the sudden arrival of the *Great Western*, fifteen days out from Bristol and with 200 tons of coal spare. The *Sirius* had got there first, by bold and daring seamanship, but the *Great Western*, overhauling her at two knots, had shown that henceforth steam navigation could be regarded as a matter of mere routine. Indeed in the next eight years she made no less than 67 crossings, including one eastbound run of twelve days six hours.

The 'Great Britain'

Never content, Brunel was planning a second and even larger ship before the *Great Western* had even completed her second voyage. This one was to be of iron, with a screw-propeller and 3,270 tons gross, an unprecedented size. In July 1839 construction of the *Great Britain* was begun. Completed four years later she was named and launched by the Prince Consort, in fact being floated out of her specially built dry dock. Her removal to the open sea presented a few problems. Being 289 feet long and 51 feet broad she was too long and too wide for the lock chambers of Bristol Dock so that it was necessary for the masonry of the sidewalls of the locks to be removed before she could squeeze through on the high tide, an operation which was accomplished with only inches and minutes to spare. Fitting out was not completed until the summer of 1846 but her maiden voyage was remarkably smooth and on 22nd September 1846 she left Liverpool on her fifth voyage with a record complement of 180 passengers. Within a few hours the ship ran aground in

A hand-coloured lithograph, published by George Davey and Son, showing the 'Great Britain' being floated out of the Great Western dry dock, Bristol, on 19th July 1843.

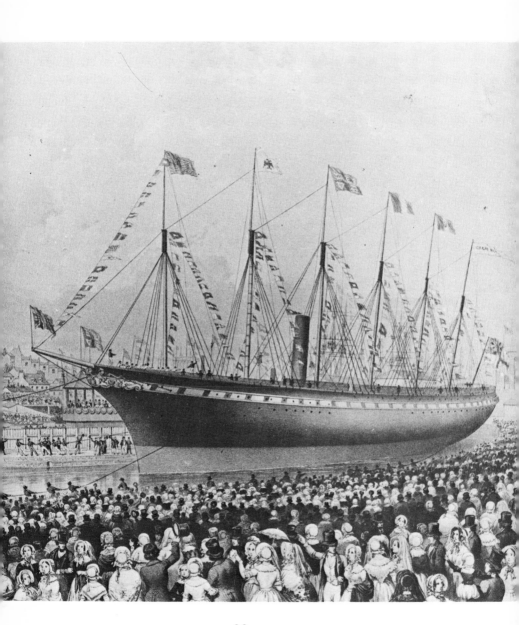

darkness. Pandemonium broke out but no lives were lost as the vessel was securely stuck on the Irish coast, not, as her captain thought, off the Isle of Man.

The immense strength of the *Great Britain* saved her where any wooden ship would have broken up, but it was not the sort of proving that Brunel was looking for; though when salvage experts failed to refloat the holed vessel, he came to her rescue. Exasperated that 'the finest ship in the world', as he modestly termed his creation, should be left 'lying, like a useless saucepan

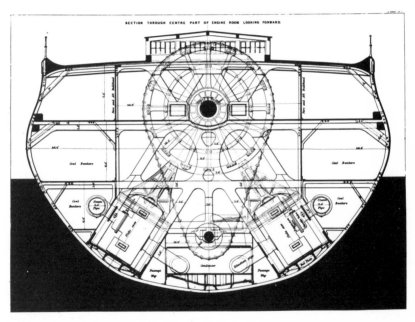

A cross section of the engine room of the 'Great Britain'.

kicking about on the most exposed shore you can imagine', he ordered the ship to be protected from the winter gales by 'a mass of large, strong faggots lashed together, skewered together with iron rods, weighted down with iron, sandbags etc., wrapping the whole round with chains, just like a huge poultice'.

Successfully reclaimed the next spring, the *Great Britain,* having bankrupted the Great Western Steamship Company, was sold in December 1850 to Gibbs, Bright and Company and put on the Australia run, carrying emigrants and bullion, and, during the Crimean War and Indian Mutiny, acting as a troopship. Her sister the *Great Western* was successfully employed on the West Indies mail run before being broken up in 1856. The *Great Britain,* having served 23 years on the Australia run was laid up in 1875 and remodelled in 1882 as a sailing ship, with her iron hull sheathed in wood. Two successful voyages to San Francisco were followed by a disastrous venture to Panama, in which she was battered and partly dismasted by a terrific gale off the Horn. Despite this she reached the safety of the Falkland Islands, where she was condemned and bought as a storage hulk for wood and coal, an ignoble duty she performed until 1937 when she was towed away and beached in Sparrow Cove, a resting place from which she was rescued in 1970. Transported, still intact, to her native Bristol, she is being lovingly restored.

The 'Great Britain' returning to her birthplace dock at Bristol on 19th July 1970. On board were H.R.H. Prince Philip, and Mr. Jack Hayward who gave £150,000 for the ship's salvage.

Crimean incident

Brunel rarely allowed political events to interfere with his professional life, though he gave evidence to a number of official commissions on railways, convinced the Admiralty of the value of the screw propeller (no mean achievement), dutifully enrolled as a special constable for the expected Chartist revolution in 1848, and served with distinction on various committees connected with the Great Exhibition of 1851. The Crimean War, however, and the official incompetence it revealed, filled him with indignation and he invented a 'floating siege gun' (to be transported by a mother ship with opening bows) to break the deadlock at Sevastopol. The idea gained the enthusiastic support of General Burgoyne, the leading military engineer of the day, but the Admiralty managed to do nothing about it quite successfully until Brunel gave up in disgust.

His remarkable talents were, however, soon brought into play at the request of his brother-in-law, Sir Benjamin Hawes, Permanent Under Secretary at the War Office. Stung by Florence Nightingale's bitter complaints about the insanitary conditions at the military hospital of Scutari, he invited Brunel to design a prefabricated hospital to be shipped out and assembled on the spot. The request came on 16th February 1855; by 5th March Brunel was explaining to Hawes the principles which have guided the layout of this type of temporary building ever since. The hospital would consist of standard units of two wards, taking 24 patients each and complete with its own water closets, outhouses and nurses' rooms. Baths, wash basins and ventilator fans were all included as well as a self-contained drainage system. Units could, of course, be joined to one another according to the size and nature of the ground. Everything was foreseen. There were even printed handbills to explain to the soldiers how they were to use a water closet. By 12th July a hospital with 300 beds had been erected by the eighteen-man gang sent out from England and by December it was equipped with its full quota of a thousand beds. In the short time it was operating 1,500 men passed through it. Only fifty of these patients died. Brunel dismissed the project as 'just a sober exercise of common sense'.

Leviathan—the 'Great Eastern'

In the course of his work for the Great Exhibition Brunel was perforce thrown into contact with John Scott Russell, cofounder of the Exhibition, Fellow of the Royal Society and Vice-President of both the Institution of Civil Engineers and the Institute of Naval Architects. By repute he was the most brilliant marine engineer of his day and Brunel, whose perceptive judgement failed him most signally on this occasion, invited him to join him in the most stupendous project of his career—a plan for a steamship so vast it would be able to voyage right round the world without refuelling. It would travel non-stop to

John Scott Russell (1808—1882), Brunel's ill-chosen partner in the building of the 'Great Eastern'.

Australia or take a whole year's exports to India, disgorge and return with a vast cargo of cotton and spices. Scott Russell was soon fired with Brunel's enthusiasm and between them they raised £120,000 through their many personal contacts in business and engineering. Railway contractors like Brassey and Peto took a thousand shares apiece and Brunel sank a very considerable part of his own fortune in the venture. As he put it himself, 'I never embarked in any one thing to which I have so entirely devoted myself, and to which I have devoted so much time, thought and labour, on the success of which I have staked so much reputation and to which I have so largely committed myself and those who were disposed to place faith in me'.

Scott Russell's own yard was awarded the contract to build the 'great ship', although his tender of £377,200 was a third lower than Brunel's estimate and vague in its terms. Brunel, who had a high opinion of both Russell's managerial and technical competence, appears to have accepted this. Experience was to reveal that the figures quoted were little better than guesses. No one could foresee the problems to be encountered in the production of castings and forgings of the required dimensions, but all this lay in the troubled future and in the meantime Brunel busied himself with constant re-styling to eliminate unnecessary weight and expense. For the sake of economy he had designed the whole ship with only two thicknesses of plate and two sizes of angle iron.

The site chosen for construction was the Napier Yard, London, adjoining Scott Russell's (whose workshops would thus not be impeded) and linked with it by a special railway. The ground was prepared for its great burden by studding it with deep wooden piles, the top four feet of each protruding from the ground and angled to enable the completed hull to be launched sideways, the only way in which a ship of such size could be controlled or prevented from crashing into the river bank opposite.

Even before construction began, relationships between Brunel and Scott Russell were soured by a long newspaper article which was full of detailed information and credited Brunel with the mere idea of a 'great ship', Scott Russell with its design and construction. Although the source of information was unnamed, it was clear to Brunel that his colleague was responsible, though he avoided a confrontation over the issue for the sake of the ship. Then, on New Year's Day 1855, Russell informed him, for the first time, that Martin's Bank would allow him no further credit. Brunel suspected nothing and made arrangements for Russell to be paid by instalments. Relations between them deteriorated rapidly, however, as Russell began to deliberately withhold co-operation, refusing to supply Brunel with data and progress reports. Brunel, although impatient, contented himself with polite requests, until August when, with the money almost gone and much work still unstarted, he sent Russell the abrupt

The 'Great Eastern' under construction at the Napier Yard, London. It was the largest ship that had ever been built, designed to carry a whole year's exports to India in one trip.

warning that 'unless . . . on *Monday next* we are busy as ants at 10 different places now untouched, I give it up'. Russell's reply was a demand for an extra £37,673.18s. to complete the ship and a long haggle over alterations began. Brunel, now exasperated, ended one letter—'I wish you *were* my obedient servant, I should begin by a little flogging . . .'

Russell, by this time, had quite lost interest in the great ship project and had not only taken on other orders and was diverting labour to them, he was even using the Napier Yard as if it were his own and obstructing work on the stern of the great ship. Having been paid a total of £292,295 Russell had completed little more than a quarter of the work on the hull. When the Eastern Steam Navigation Company reluctantly moved to take over construction itself, Russell 'discharged all the men connected with the works of the ship' and revealed that he had even leased out the ground on which the bow portion of the ship now rested. Russell, however, had no assets which Brunel and his partners could distrain, had complete possession of all detail drawings, and also the loyalty of Dixon and Hepworth, two of the key foremen who had supervised the work from its beginning. Arrangements to buy off the latter and obtain permission to go on using their own yard, were made at the cost of much time and expense, and every effort made to speed completion. Brunel's misgivings were overridden by the need for haste but even so the launch deadline of October 1857 was not fulfilled and it was not until 3rd November that the first attempt was made to launch the 12,000 ton hull.

Brunel was fully aware that the launch was the greatest technical challenge of his career and he had devoted months of painstaking experiment and calculation towards solving the problem of moving this vast mass without letting it get out of control. One slip could wreck the whole project for ever, but he had the recent experience of the Saltash bridge to guide him and was prepared to proceed slowly and cautiously. Unfortunately, the popular press was not, and, ignoring Brunel's warning that the proceedings would be quite unspectacular, drummed up a vast crowd to watch the attempt. In addition the directors, without telling Brunel, had sold 3,000 tickets giving admission to the yard, in an attempt to raise some cash. When Brunel arrived to direct operations he found, to his horror, the Isle of Dogs decked out in holiday style and a milling chaos of visitors around the great ship itself. The total silence which had made the Saltash operation such a success was quite impossible and the events of 3rd November became a grotesque parody of that rehearsed and disciplined exercise. In the event the bow section moved three feet, the stern four, and the sightseers struggled home through the damp, foggy evening consoled only by the horrible death of an aged labourer whose legs had been smashed by a runaway winch.

On 19th November a second attempt was made, using two additional hydraulic rams. No progress was made. On the 28th 14 feet of movement was achieved and by the 30th it had moved a total of 33 feet 6 inches from its original resting-place, at a considerable cost in broken chains and winches. Even the hydraulic rams burst under the pressure and had to be replaced. On 3rd and 4th December the ship was moved a further

Right: at the launching of the 'Great Eastern'; from left to right, John Scott Russell, Henry Wakefield, Brunel and Solomon Tredwell.

44 feet and then attempts were suspended for a month while Brunel assembled a motley collection of 36 hydraulic presses and rams, giving a total thrust of 4,500 tons. Had he been allowed the specially designed equipment he originally wanted these stop-gap measures would have been unnecessary. Between 5th and 14th January the hull was eased to the bottom of the slipway, so that she became partially waterborne at high tide. By waiting another fortnight Brunel caught the full spring tide for his next effort and at 1.42 p.m. on 31st January 1858 'Leviathan' as it had now been christened, was afloat and, by evening, secured in her new moorings at Deptford. When Brunel came down the ship's side to a spontaneous ovation from the workmen he was a broken man. Months of labour and anxiety had taken their toll and the last sixty-hour effort to float 'Leviathan' had brought him to the point of exhaustion.

From May to September he travelled the Continent, theoretically 'resting', in fact working on designs for the Eastern Bengal Railway. He returned to participate in the sale of the uncompleted ship to a new 'Great Ship Company', as the Eastern Steam Company was wound up. Having prepared specifications for the ship's fittings he was advised again to take a holiday for the sake of his health. Nephritis, a kidney disease, had been diagnosed and he was recommended to a warm climate. On Christmas Day he dined in Cairo with his old friend Robert Stephenson and then proceeded, to the delight of his son and the dismay of his wife, to hire local labour and modify a wooden date boat to ascend the rapids above Assuan. When he returned to England in May 1859 it was to find that the directors of the new company had awarded the contract for fitting out the ship to—Scott Russell! Wranglings and disputes had already begun and Brunel, deaf to the orders of his doctors and the pleas of his friends, summoned up his last reserves of energy and took charge of operations in person. He knew that he was a dying man. His last ambition was to see the *Great Eastern* as he, and by now everyone else, called her, afloat and steaming majestically through the Channel to the open sea. His decision to squander his last strength on getting her to sea was, therefore, a deliberate calculation, and a tragic one.

Scott Russell was once more in fine form, back in control of his own company and at a celebration dinner on 8th August he made a long speech of self-congratulation and in the guide-book about the ship which was sold to visitors, it was claimed that

The 'Great Eastern' was finally launched on 31st January 1858, and in September 1859 underwent its first sea trials.

'the merit of the constructions of the ship and her successful completion is owing entirely to the untiring energy and skill of Mr. Scott Russell'. Brunel, meanwhile, was limping around the Napier Yard with the aid of a stick, still rattling out orders, checking and supervising, to make the *Great Eastern* ready for her first sea trials on 7th September. Illness cheated him of his final reward, however, and on 5th September he was paralysed by a stroke and carried from his ship for the last time. But the ship did meet its schedule and by the 8th, when she lay at the Nore, he was up dictating letters and instructions.

ISAMBARD KINGDOM BRUNEL
CIVIL ENGINEER
BORN 1806 DIED 1859

This statue of Brunel by Baron Marochetti stands on the Victoria Embankment, London, at the corner of Temple Place, opposite Strand Lane.

On the morning of 9th September the *Great Eastern* steamed past the Nore Light at 13 knots, gliding disdainfully through the choppy seas, to the amazement and wonder of her passengers. Meanwhile, below decks, disaster lay in wait. Around each of the five funnels was a six-inch water jacket, designed to heat the boiler feedwater and simultaneously prevent the saloon, through which glass-cased funnels passed, from becoming overheated. In the haste of preparation someone forgot to remove two temporary stopcocks fitted to the heaters on the leading and second funnel and as the great leviathan gradually worked up speed so the pressure in these heaters rose to danger level. Mercifully the saloon was deserted when the explosion occurred at 6.5 a.m. The *Times* correspondent, at the bow of the ship, described the effect—'The forward part of the deck appeared to spring like a mine, blowing the funnel up into the air. There was a confused roar amid which came the awful crash of timber and iron mingled together in frightful uproar and then all was hidden in a rush of steam. Blinded and almost stunned by the overwhelming concussion, those on the bridge stood motionless in white vapour . . . glass, giltwork, saloon ornaments and pieces of wood . . . began to fall like rain in all directions.' The men in the boiler room were hideously scalded, their bodies boiled quite white, their flesh falling away at the touch. One stumbled to the deck and threw himself screaming over the rail to be battered to pieces by the huge paddle wheel. Five more engineers died but not one passenger was harmed. Brunel's brilliant cellular construction confined the explosion entirely to the grand saloon. In the adjoining library not a book was out of place, not a mirror cracked.

Brunel's death

When the news reached Brunel it came as the final blow, a stroke too cruel for even his great spirit. Cheated of his hour of triumph he closed his eyes for the last time on Thursday 15th September.

Daniel Gooch, his closest friend, said of him: 'by his death the greatest of England's engineers was lost, the man with the greatest originality of thought and power of execution, bold in his plans but right. The commercial world thought him extravagant; but although he was so, things are not done by those who sit down and count the cost of every thought and act.'

PRINCIPAL EVENTS OF BRUNEL'S LIFE

1806   Brunel born

1820   Brunel goes to Caen
1821
1822   Brunel starts to work for his father
1823
1824
1825   Work on the Thames Tunnel begins
1826
1827   Tunnel flooded
1828   Tunnel abandoned
1829
1830   Brunel wins Avon bridge competition. Bristol Riots. *Liverpool and Manchester Railway opened*
1831
1832   *First Reform Bill*
1833   Brunel appointed engineer to Great Western Railway
1834
1835
1836   Brunel marries
1837   *Victoria becomes Queen*
1838   First section of Great Western Railway opened. *Great Western* launched
1839
1840
1841   GWR completed
1842   *Queen Victoria travels by train for the first time–on the GWR*
1843
1844
1845
1846   Maiden voyage and shipwreck of the *Great Britain*. Battle of the gauges
1847
1848   Atmospheric railway runs in Devon
1849
1850
1851   *Great Exhibition*
1852
1853
1854   *Crimean War begins.* Saltash bridge begun

Further reading

1855   Brunel designs hospital for the Crimea
1856   *Crimean War ends*
1857
1858   *Great Eastern* launched
1859   Saltash bridge completed. *Great Eastern* sails. Brunel dies.

FURTHER READING

The standard biography by L. T. C. Rolt, *Isambard Kingdom Brunel,* is available in a Penguin paperback edition. The same author's *Victorian Engineering* provides a useful context and is also available in Penguin. More detailed information can be found in:

Pugsley, Sir A. (editor). *The Works of Isambard Kingdom Brunel: An Engineering Appreciation.* 1976.

St John Thomas, D. *A Regional History of Railways in Great Britain: Volume 1, The West Country.* Fifth edition, 1981.

Body, Geoffrey. *Railways of the Western Region.* PSL Railway Guides, 1983.

SEEING BRUNEL'S WORK

A certain amount can be seen in London, where there are models of the SS *Great Britain* (National Maritime Museum, Romney Road, Greenwich, London SE10 9NF; telephone 01-858 4422) and of the *Great Western* and *Great Eastern* (Science Museum, Exhibition Road, South Kensington, London SW7 2DD; telephone 01-589 3456), but most notably the pumping station for the Rotherhithe Tunnel (at the Rotherhithe end) and Paddington Station itself, where there is also a statue of Brunel.

The real enthusiast must go west. The Great Western Railway Museum, Faringdon Road, Swindon, Wiltshire (telephone Swindon [0793] 26161, extension 3131) has its own Brunel Room, exhibiting not only photographs and engineering drawings but also the tools of his trade, including his drawing board. Bristol has not only the magnificent Clifton Suspension Bridge but also the brilliantly restored SS *Great Britain,* Great Western Dock, Gas Ferry Road, Bristol BS1 6TY (telephone Bristol [0272] 20680). Brunel's original station at Bristol Temple Meads, opened in 1840 but superseded by the present station in the 1860s, is being restored by the Brunel Engineering Centre Trust as an exhibition centre. The work will take until 1990 but in the meantime visitors will be shown round if they apply in advance to the Trust at Bristol Old Station, Temple Gate, Bristol BS1 6QQ; telephone Bristol [0272] 292688.

Further west still one can see the Royal Albert Bridge over the Tamar at Saltash,Cornwall, the pumping house for the atmospheric railway at Starcross, near Dawlish, Devon, and, at Exeter Maritime Museum, The Quay, Exeter, Devon EX2 4AN (telephone Exeter [0392] 58075), the *Bertha,* designed by Brunel in 1844 and now the world's oldest working steamboat.

INDEX

Figures in italic refer to the page numbers of the illustrations